THE ORIGIN OF THE UNIVERSE

Understanding the Universe

Astronomy Book | Science Grade 8 | Children's Astronomy & Space Books

First Edition, 2019

Published in the United States by Speedy Publishing LLC, 40 E Main Street, Newark, Delaware 19711 USA.

© 2019 Baby Professor Books, an imprint of Speedy Publishing LLC

All rights reserved.

Without limiting the rights under the copyright reserved above, no part of this publication may be reproduced, stored in or introduced into a retrieval system, or transmitted, in any form, or by any means (electronic, mechanical, photocopying, recording, or otherwise), without the prior written permission of the copyright owner.

All images in this book have been reproduced with the knowledge and prior consent of the artists concerned, and no responsibility is accepted by producer, publisher, or printer for any infringement of copyright or otherwise arising from the contents of this publication.

Baby Professor Books are available at special discounts when purchased in bulk for industrial and sales-promotional use. For details contact our Special Sales Team at Speedy Publishing LLC, 40 E Main Street, Newark, Delaware 19711 USA. Telephone (888) 248-4521 Fax: (210) 519-4043. www.speedybookstore.com

10 9 8 7 6 * 5 4 3 2 1

Print Edition: 9781541949706
Digital Edition: 9781541951501

See the world in pictures. Build your knowledge in style.
https://www.speedypublishing.com/

CONTENTS

Dimensions of the Universe 7
Age of the Universe 17
Steady State Theory 25
Big Bang Theory 29
Matter Forms 39
Cosmic Microwave Background 45
The First Elements 51
Formation of Stars 55
Formation of Planets 65
Summary 70

The most colorful view of the universe captured by a space telescope

Our universe is the sum of everything that exists. Matter, energy, space-time, and everything that is yet to be discovered. The universe is also known as the cosmos. This book will help you to understand how the universe may have begun, as well as how its age and dimensions can be determined.

DIMENSIONS OF THE UNIVERSE

The he universe can be divided into two parts: the observable universe and the non-observable universe.

> The observable universe includes all things that we can see from Earth

The observable universe includes all things that we can see from Earth. We can only see stars once their light has reached earth. Therefore, the light we see today from the most distant stars is about 13.8 billion years old.

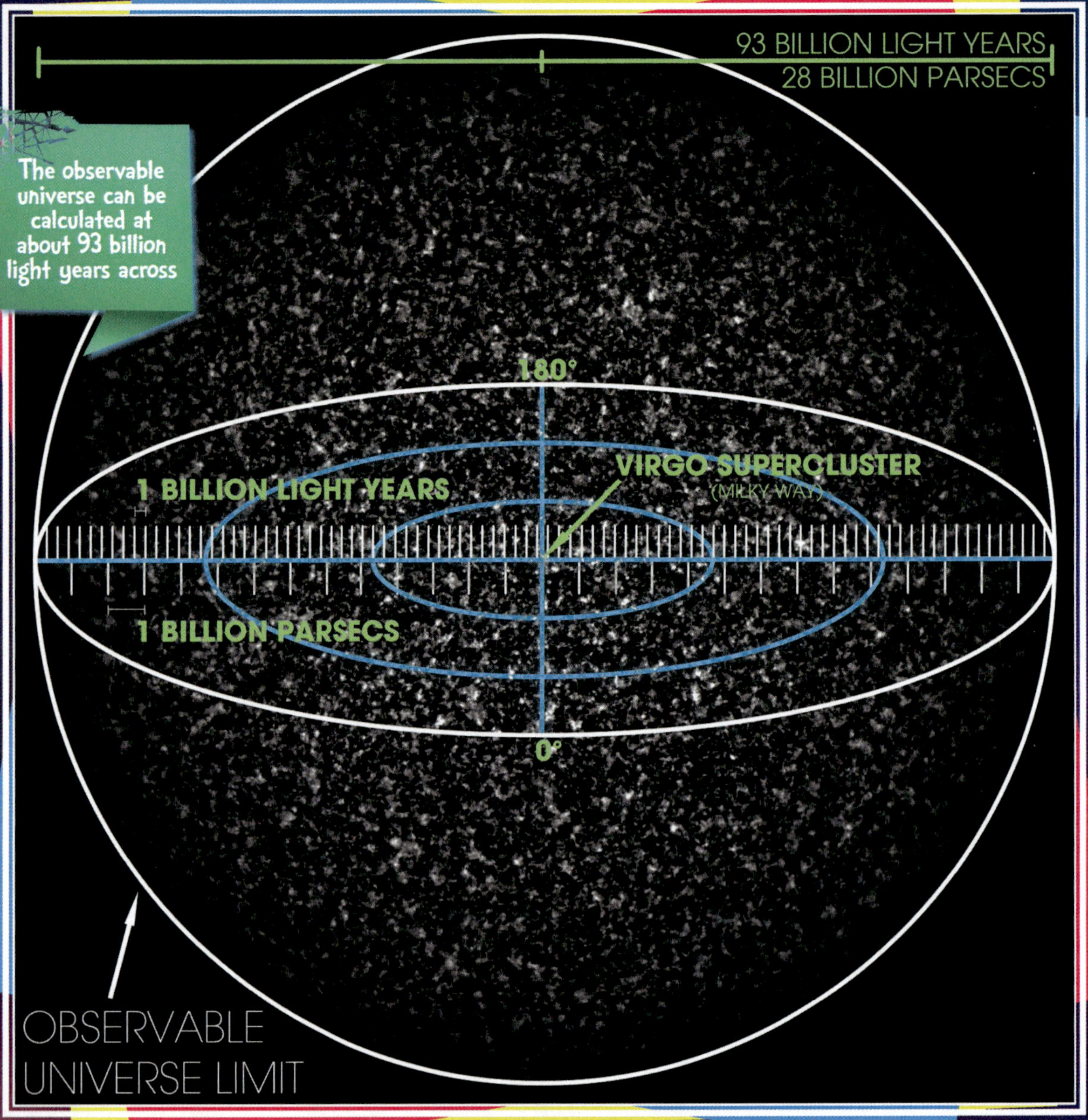

93 BILLION LIGHT YEARS
28 BILLION PARSECS

The observable universe can be calculated at about 93 billion light years across

180°

1 BILLION LIGHT YEARS

VIRGO SUPERCLUSTER
(MILKY WAY)

1 BILLION PARSECS

0°

OBSERVABLE UNIVERSE LIMIT

Since we can see 13.8 billion light years in every direction, this would make the observable universe about 27.6 light years across. However, we know that these stars are moving away from us at an accelerated rate. The most distant stars that we see today are actually much further away than they appear. Therefore, the observable universe can be calculated at about 93 billion light years across.

The speed of light is calculated to be 186,000 miles per second (299,792.458 kilometers per second). A light year is the distance light travels in one year. Therefore, one light year equals 5.879×10^{12} miles (9.461×10^{12} kilometers). The universe, as we see it today, is a sphere with a diameter roughly 5.467×10^{23} miles (8.798×10^{23} kilometers). However, the observable universe expands by about three light seconds for each second that passes.

Speed of Light = 299,792 km/sec

Earth to Moon = 1.3 Light seconds

Earth to Sun = 8 Light minutes

Earth to Mars = 12.7 Light minutes

Earth to Alpha Centauri = 4.4 Light years

Earth to far side of our MIlky Way Galaxy = 52,000 Light years

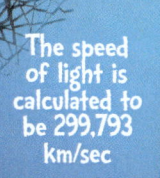

The speed of light is calculated to be 299,793 km/sec

The non-observable universe is everything that exists beyond the observable universe. We do not know how big it is now because we have no way of measuring it. However, it could, theoretically, be infinite and constantly expanding.

THE ENTIRE UNIVERSE

Boundary of the Observable Universe

The non-observable universe is everything that exists beyond the observable universe

AGE OF THE UNIVERSE

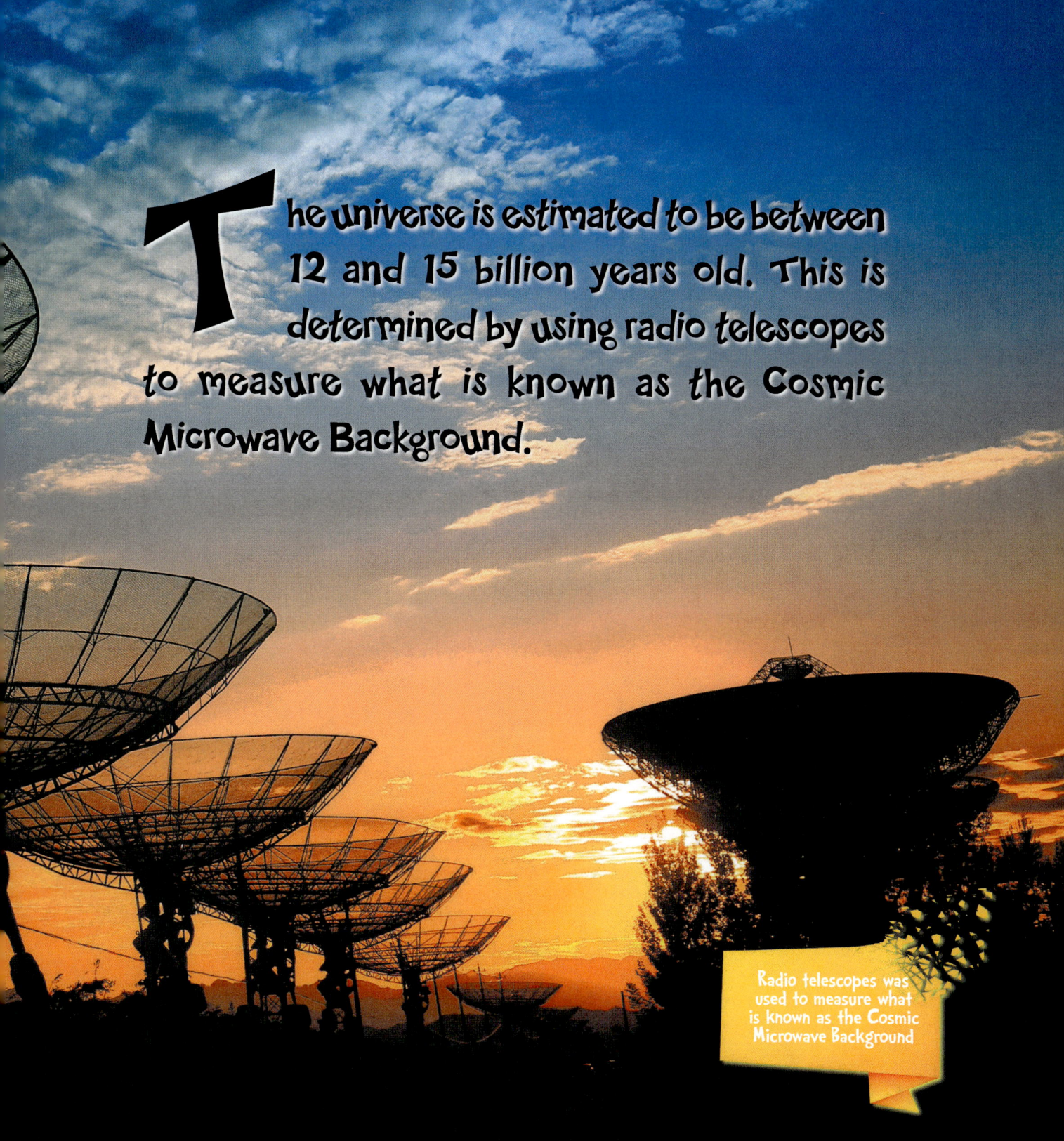

The universe is estimated to be between 12 and 15 billion years old. This is determined by using radio telescopes to measure what is known as the Cosmic Microwave Background.

Radio telescopes was used to measure what is known as the Cosmic Microwave Background

As the universe expands, the earliest particles of light continue to move outward. The Doppler Effect causes these light waves to stretch out to wavelengths longer than those in the spectrum of visible light. This early cosmic radiation is now in the form of microwaves.

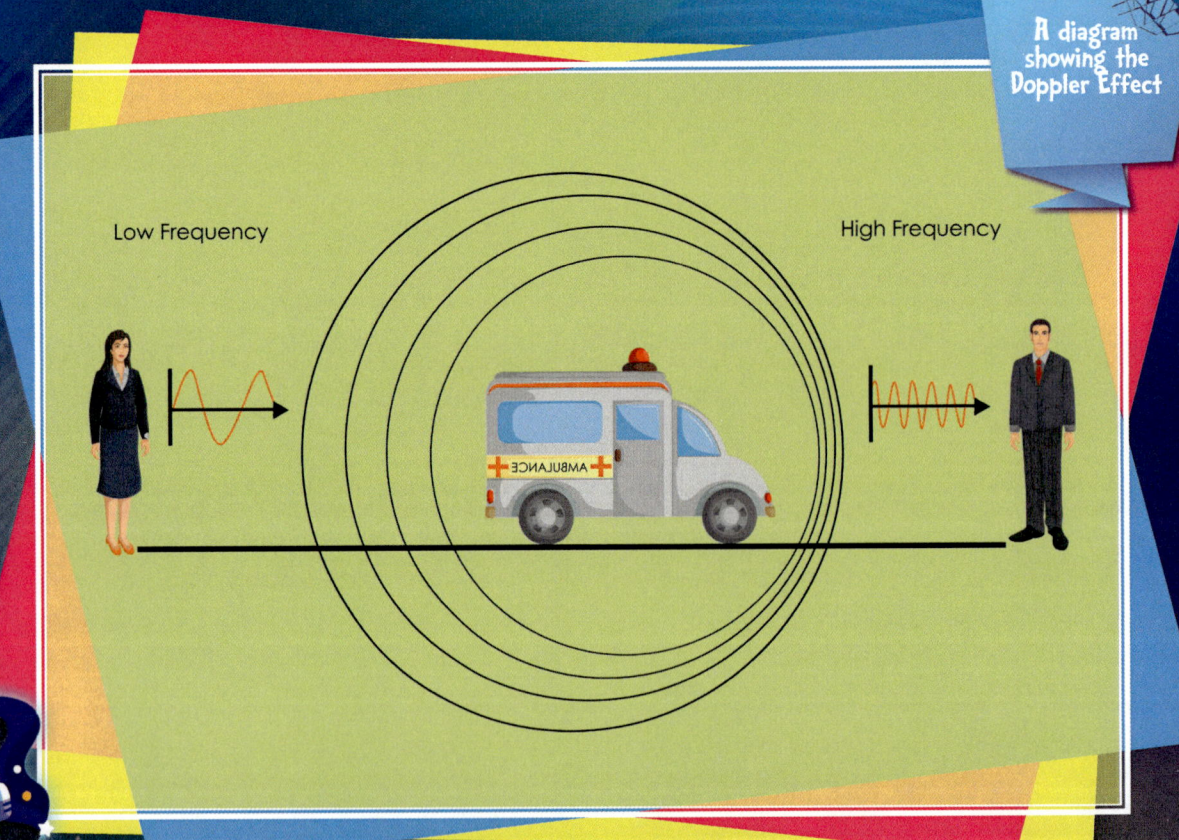

A diagram showing the Doppler Effect

COSMOLOGICAL REDSHIFT

Original Wavelength

Stretched (Redshifted) Wavelength

TIME

An illustration of a stretched and original space wavelength with earth and distant galaxy

The rate of expansion of the universe can be calculated by observing the Cosmic Microwave Background. Using this rate of expansion, we can then extrapolate backwards, or rewind, to the point where the universe started. The most recent estimation of the age of the universe is 13.77 billion years. This estimation is based on the Big Bang Theory.

STEADY STATE THEORY

A Hubble image showing an infant galaxy forming nearby

Concerning the origins of our universe, there are two predominant theories: the Steady State Theory and the Big Bang Theory.

The Steady State Theory suggests that the universe has no beginning or end. It holds that the universe is expanding but not becoming less dense. This is based on the idea that energy is always being converted into hydrogen. Hydrogen then condenses to form new stars and galaxies. These newer galaxies replace older galaxies as they move away and can no longer be seen.

BIG BANG THEORY

BIG BANG THEORY

Inflation	First Particles	First Nuclei	First Light	Dark Ages	Gravity	Antigravity
Quarks Form	Neutrons, Protons, Dark Matter form	Helium, Hydrogen form	First Atoms Form	Clumps of Matter Form	Stars and Galaxies Form	Universe Expansion Accelerates

	milliseconds 10^{-32}	milliseconds 0.01	seconds 0.01 - 200	years 380.000	years 380.000	years 300 million	years 10 billion
TIME							
SIZE	Grapefruit	0.1 - trillionth present size	1 - billionth present size	0.0009 present size	0.9 present size	0.1 present size	0.77 present size

The Big Bang Theory states that the universe began as a super-dense, superheated mass about one trillionth the size of a needle point. All of the energy and matter that exists in the universe today was condensed into this mass.

There are four types of forces that govern the universe. They are gravity, the strong nuclear force, the electromagnetic force, and the weak nuclear force. These forces were mixed together and indistinguishable in the early universe.

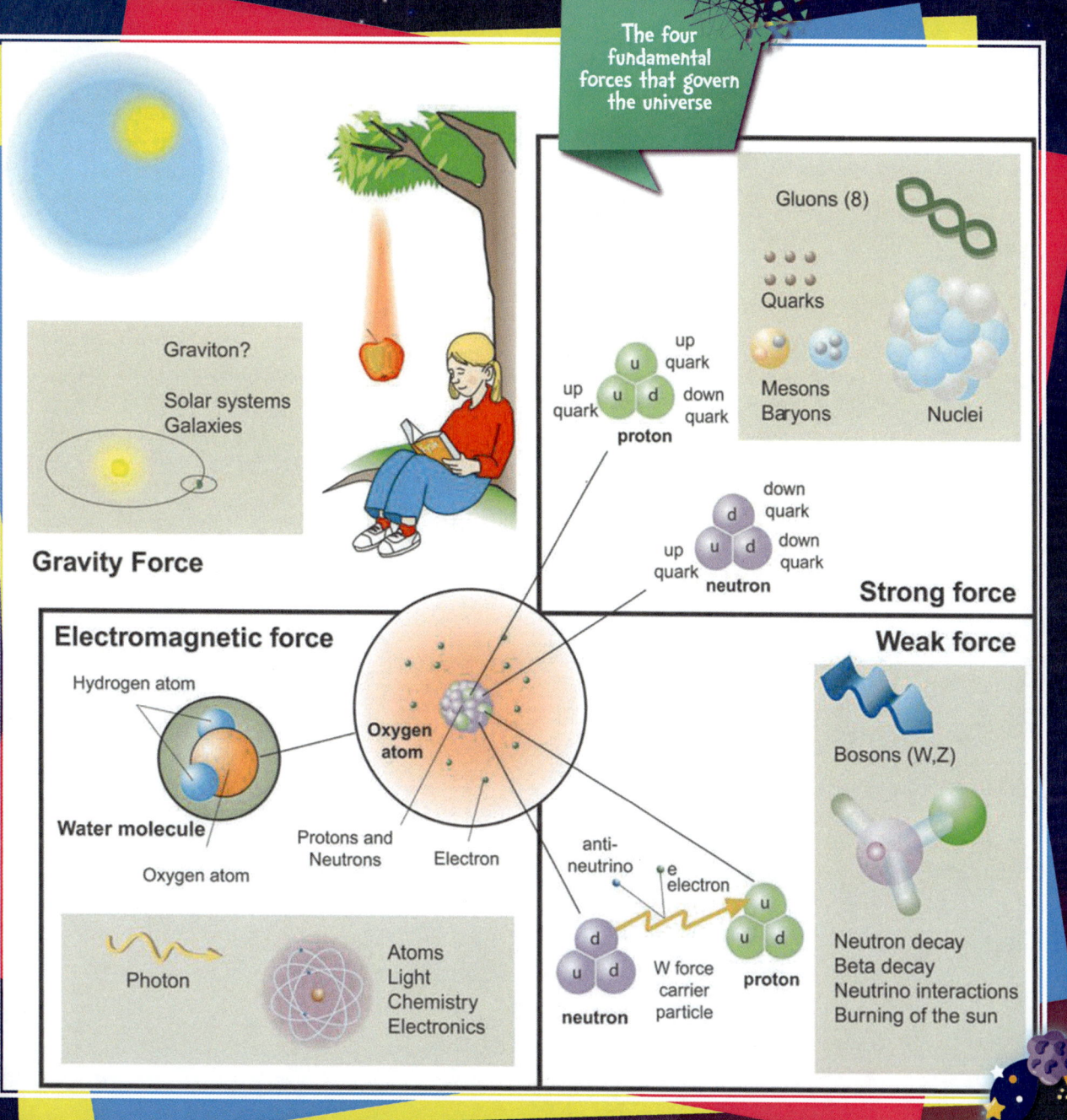

As the universe expanded and cooled, gravity separated out and became a unique force. Next, the strong nuclear force split away. This caused an incredible amount of energy to be released. Particles and energy were launched in every direction causing the universe to expand rapidly.

Generally, the Big Bang Theory states that the universe will continue to expand forever. In fact, the rate at which the universe is expanding is increasing. However, an alternate theory is that the universe will eventually stop expanding and contract, crushing everything back together. This theory holds that the universe goes through a cycle of expansion and contraction every 80 billion years.

Big Bang

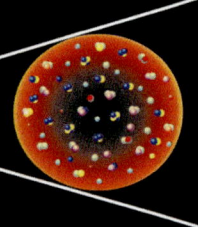

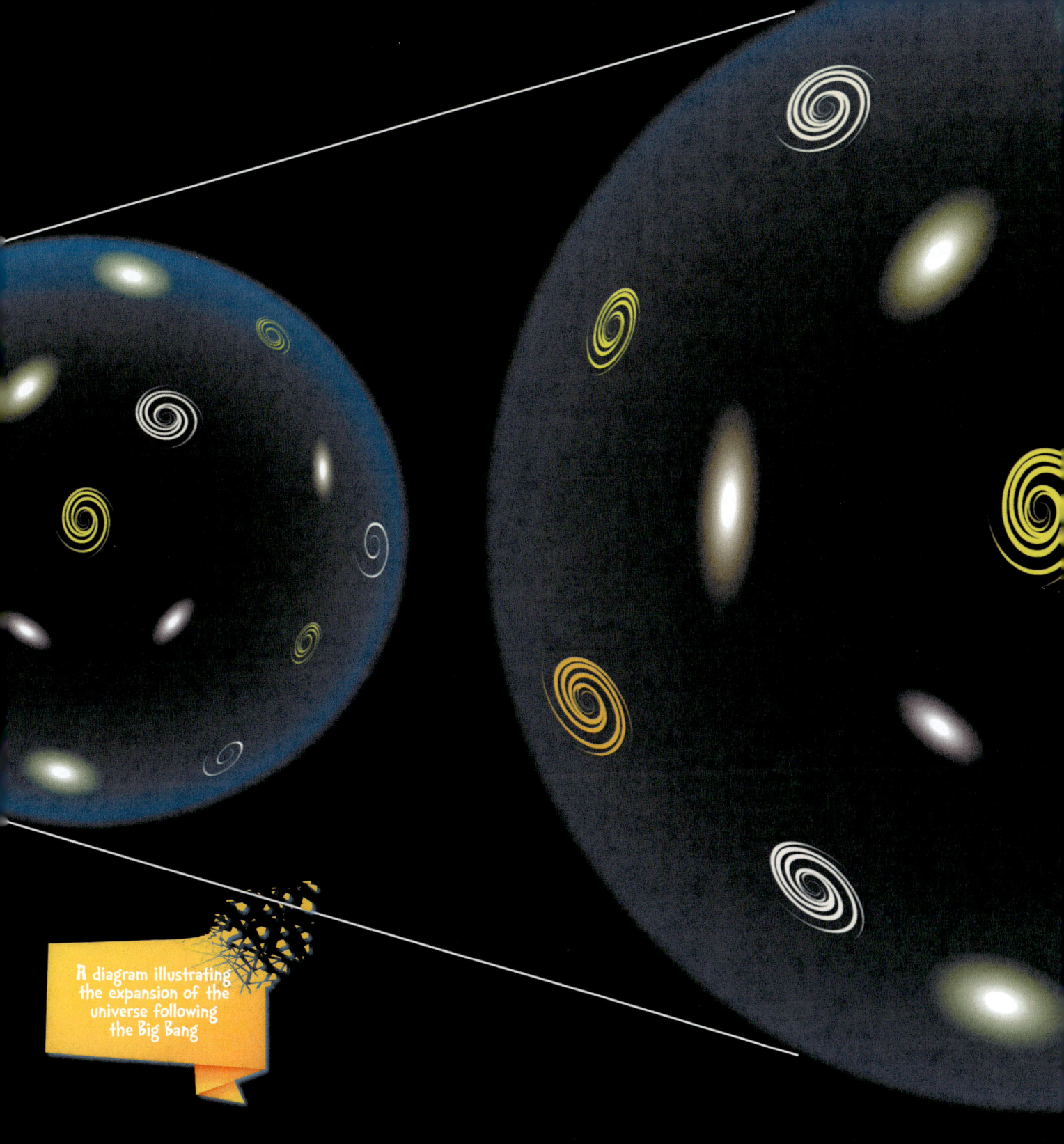

A diagram illustrating the expansion of the universe following the Big Bang

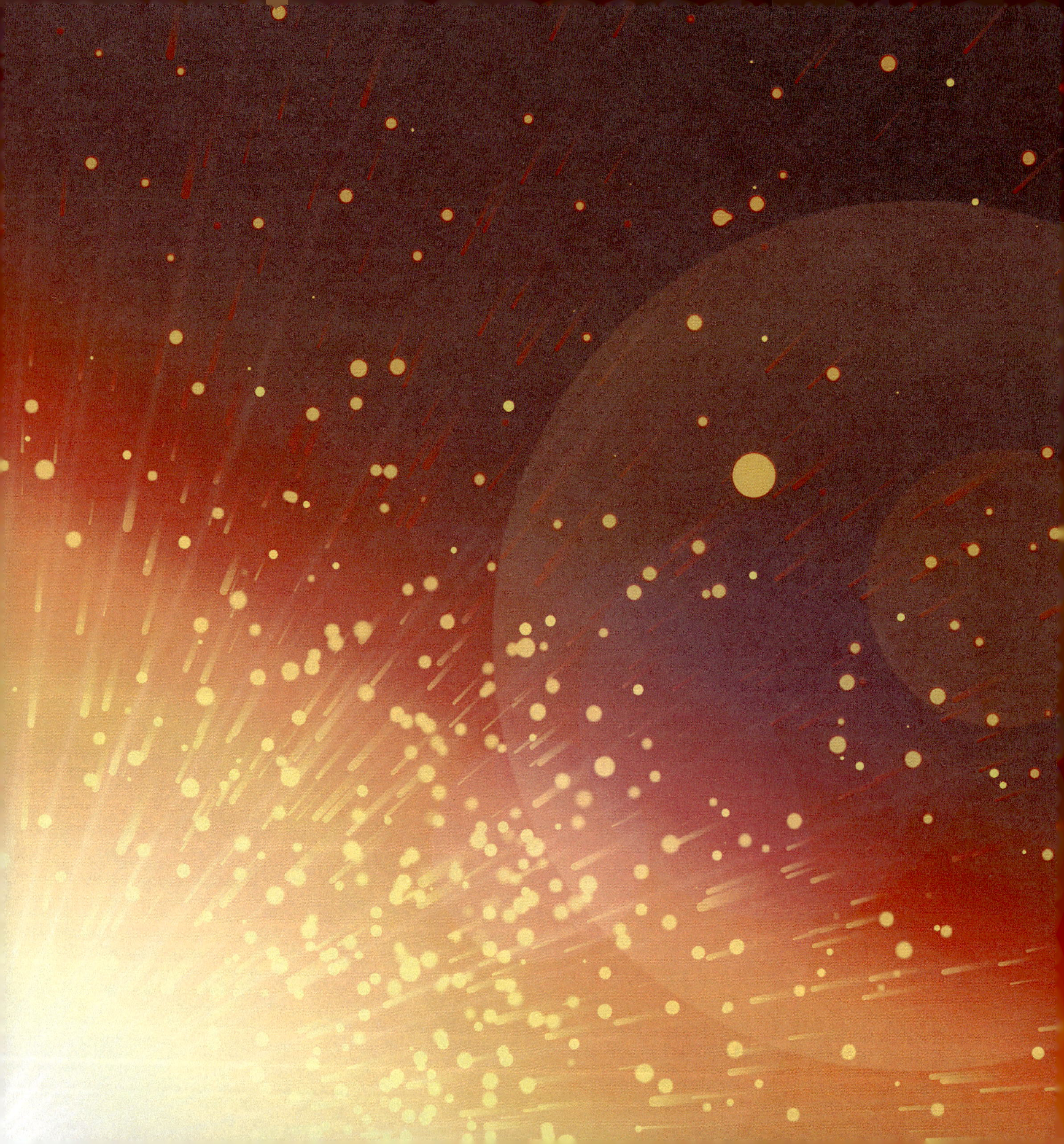

MATTER FORMS

> Photons continually transformed their energy into particles of matter and anti-matter

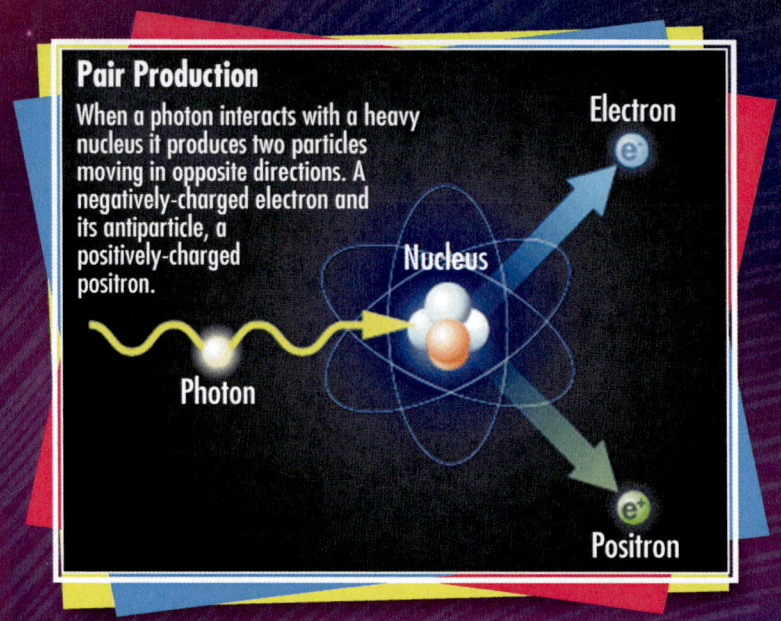

Pair Production

When a photon interacts with a heavy nucleus it produces two particles moving in opposite directions. A negatively-charged electron and its antiparticle, a positively-charged positron.

The universe was still incredibly hot. Photons continually transformed their energy into particles of matter and anti-matter. When a particle of matter came into contact with the same particle type of anti-matter, they destroyed each other and released energy. This energy was then reabsorbed into the photons, and the process began again.

However, for every one billion particles of anti-matter created, one billion and one particles of matter were created. This was a very important development that would lead to the creation of the universe as we know it today. Without that one extra particle of matter per billion, the universe would contain nothing but light.

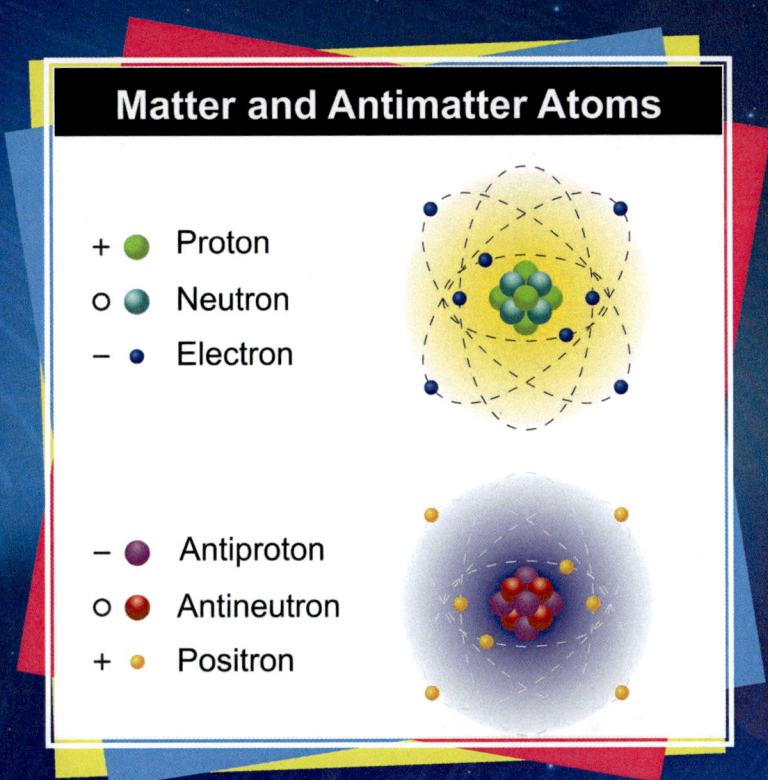

Matter and Antimatter Atoms

+ Proton
o Neutron
− Electron

− Antiproton
o Antineutron
+ Positron

For every 1 billion particles of anti-matter created, 1 billion and 1 particles of matter were created

The electromagnetic force is separated from the weak nuclear force as the universe continued to cool. Now, the four distinct forces that govern our universe were independent of each other. As the temperature dropped, photons no longer had enough energy to form new particles of matter and anti-matter. Any remaining anti-matter was annihilated by its paired matter. The universe now had a single particle of matter per billion photons.

The universe now had a single particle of matter per billion photons

BIRTH OF THE UNIVERSE

10^{-32} seconds

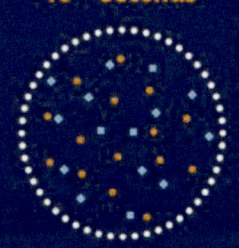

Matter and antimatter condense from pure energy. In theory, there should be equal amounts of each. Universe still less than 2m wide

10^{-32} – 10^{-6} seconds

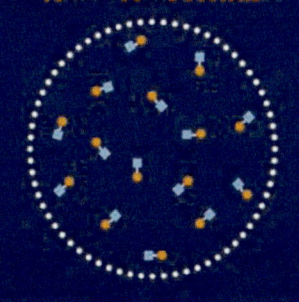

Matter and antimatter particles collide and vanish but...

10^{-36} seconds*

Universe consists of a rapidly expanding fireball of intense energy. Matter does not yet exist

...after 13.7 billion years

There's enough matter left to form the billions of galaxies of the known universe.

Beginning of time

• Universe exists as a 'gravitational singularity'

* 10^{-36} seconds as a decimal would have 36 zeroes between '0.' and '1'

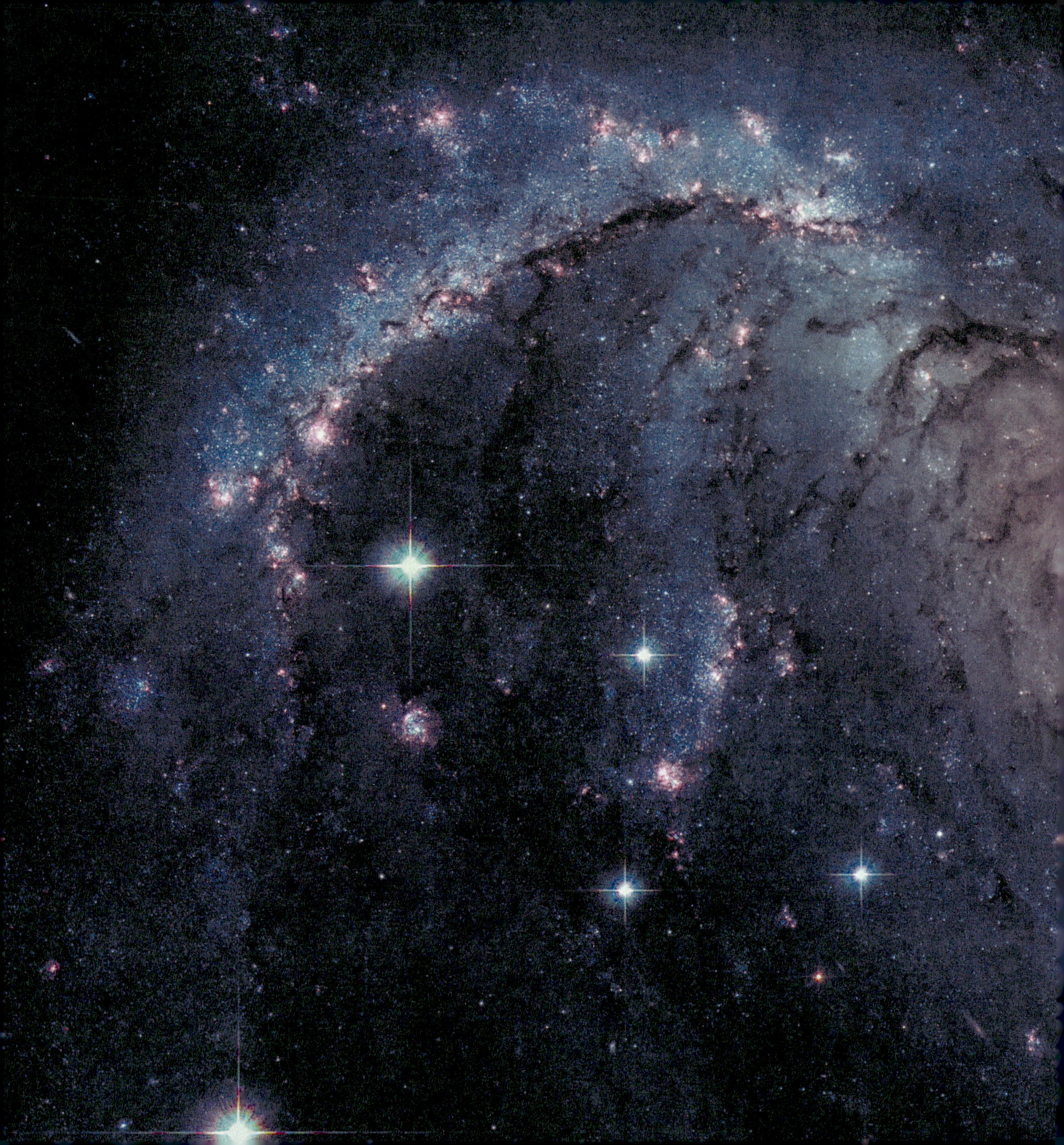

COSMIC MICROWAVE BACKGROUND

Soon, the universe had cooled enough for protons and neutrons to combine and form simple atomic nuclei. However, the universe was still too hot and energetic for negatively charged electrons to be captured by positively charged protons. Elements could not yet form.

Similarly, light could not travel in a straight line. The free electrons constantly buffeted and scattered the photons. This resulted in a sort of soup comprised of matter and energy.

ATOM

The universe had cooled enough for protons and neutrons to combine and form simple atomic nuclei

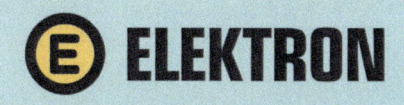

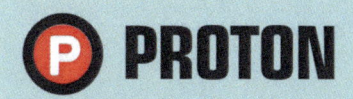

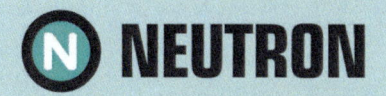

NUCLEUS

E ELEKTRON **P** PROTON **N** NEUTRON

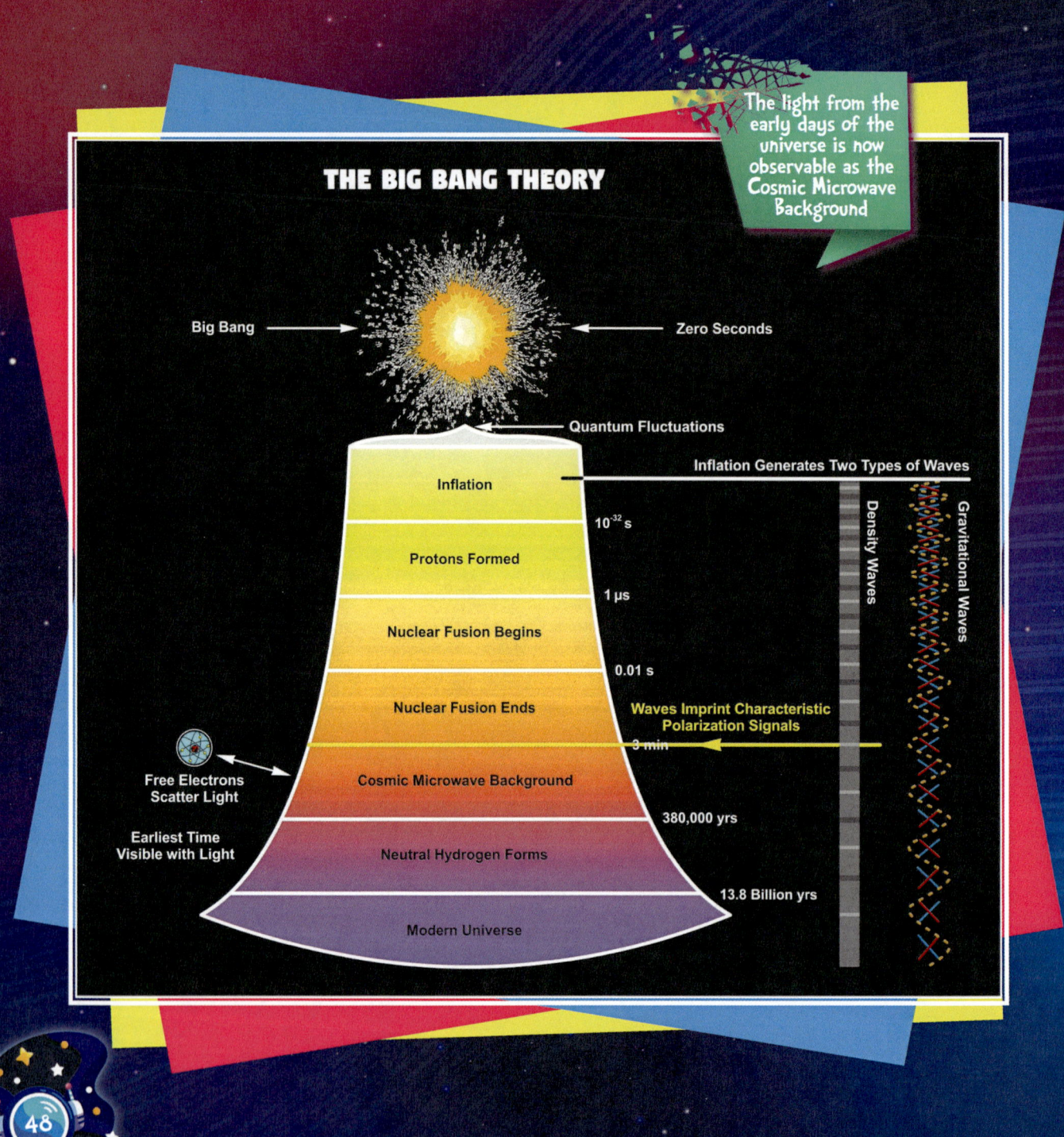

THE BIG BANG THEORY

The light from the early days of the universe is now observable as the Cosmic Microwave Background

Big Bang → ← Zero Seconds

Quantum Fluctuations

Inflation — 10^{-32} s

Protons Formed — 1 μs

Nuclear Fusion Begins — 0.01 s

Nuclear Fusion Ends

Free Electrons Scatter Light

Earliest Time Visible with Light

Cosmic Microwave Background — 3 min

Neutral Hydrogen Forms — 380,000 yrs

Modern Universe — 13.8 Billion yrs

Inflation Generates Two Types of Waves

Density Waves | Gravitational Waves

Waves Imprint Characteristic Polarization Signals

After about 380,000 years had passed, the universe was finally cool enough for the first atoms to form. With electrons safely locked into orbit around atomic nuclei, photons could travel away unhindered. This light from the early days of the universe is now observable as the Cosmic Microwave Background.

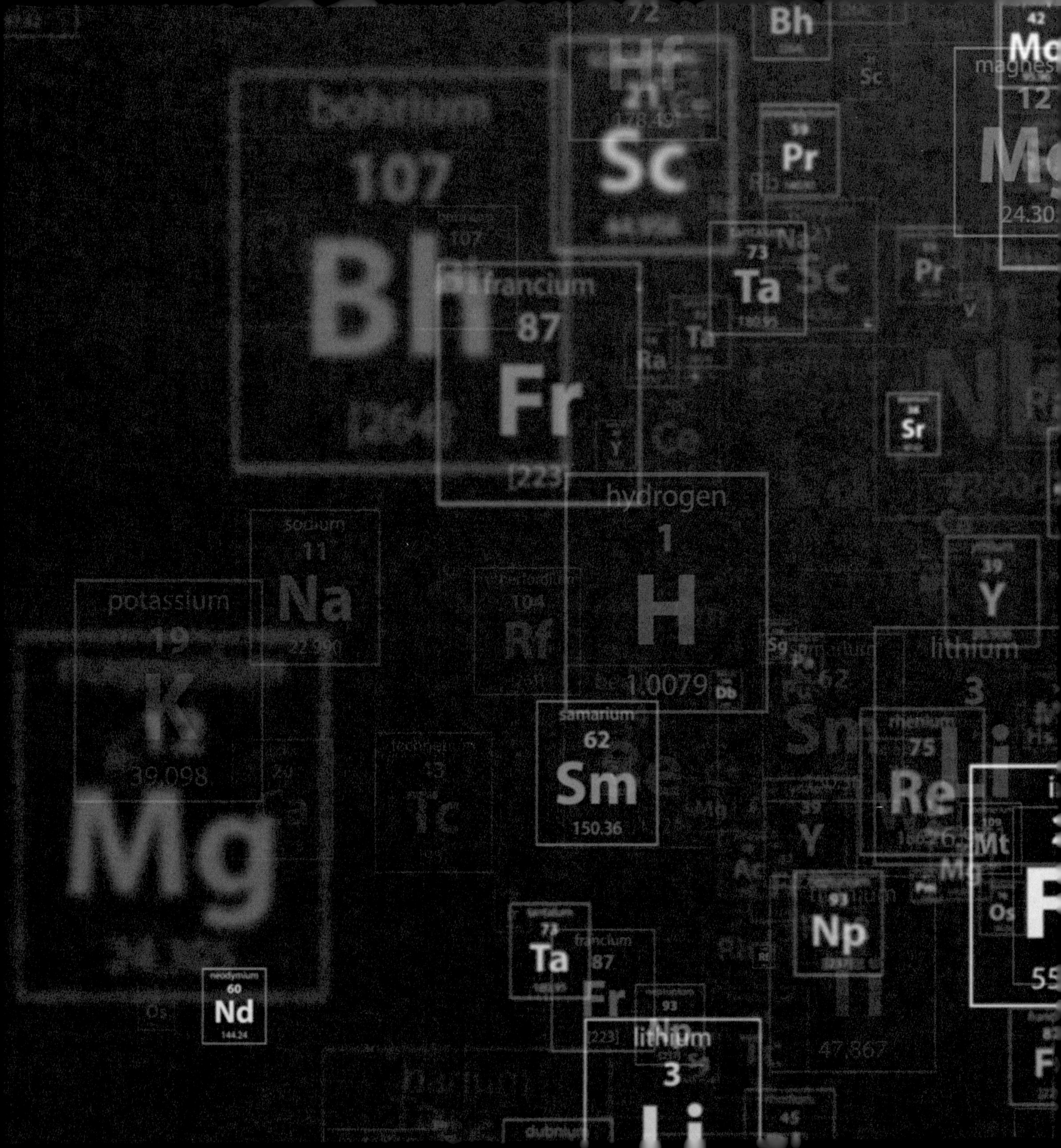

107
Bh
francium
87
Fr
[223]
Sc
73
Ta
Pr
Sr
hydrogen
1
H
1.0079
sodium
11
Na
potassium
19
K
39.098
Mg
samarium
62
Sm
150.36
39
Y
lithium
3
Li
75
Re
93
Np
neodymium
60
Nd
144.24
Mt
Os
47.867

THE FIRST ELEMENTS

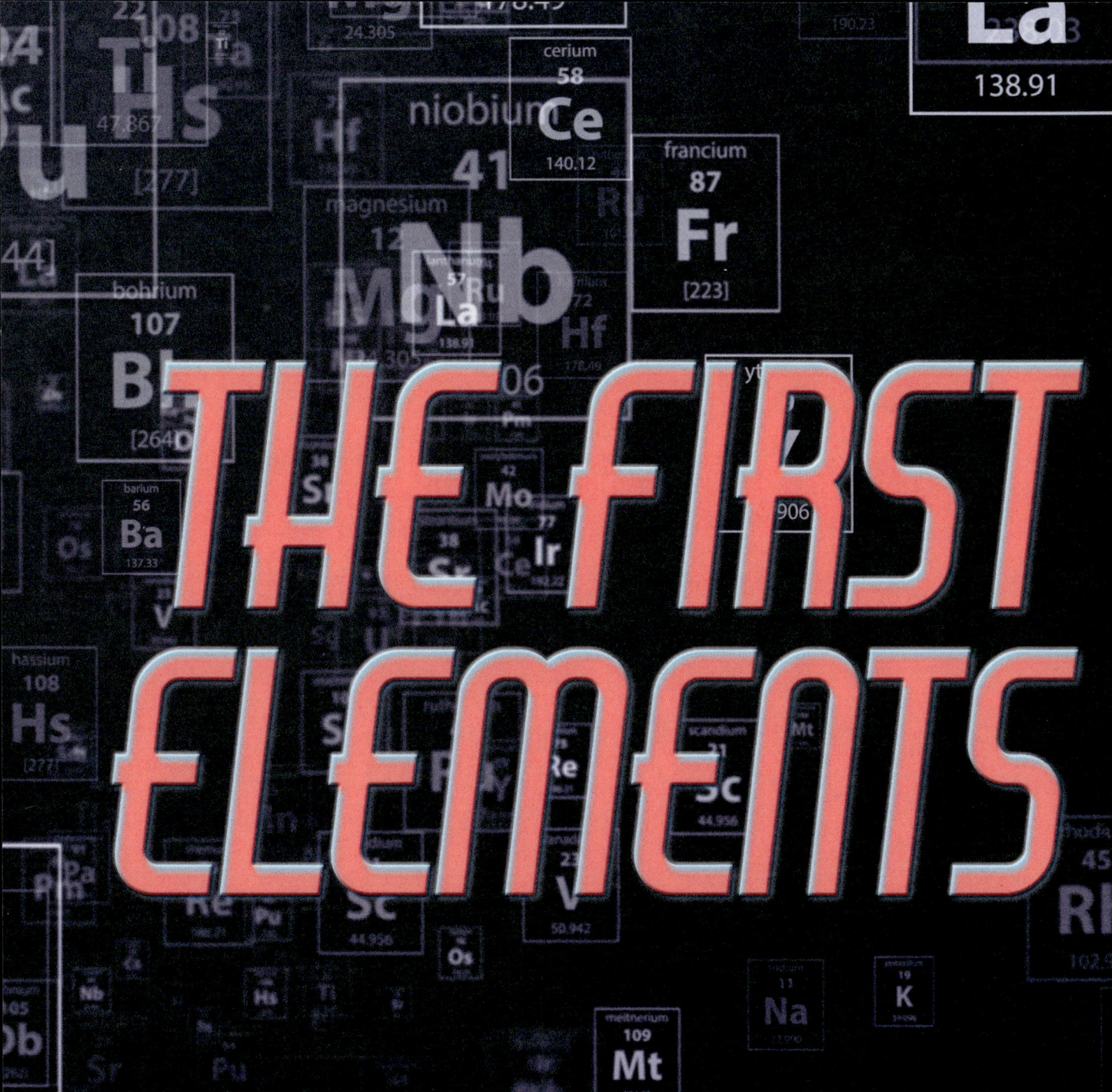

Hydrogen, helium, and lithium are the lightest elements and would have been the earliest atoms to form. Hydrogen would have made up about 75 per cent of those first atoms. The other roughly 25 per cent would have been helium, with only trace amounts of lithium forming.

Today, these elements appear in those exact same proportions, providing evidence for the Big Bang Theory.

1 Hydrogen — H

Atomic mass: 1.008
Electron configuration: 1

2 Helium — He

Atomic mass: 4.0026
Electron configuration: 2

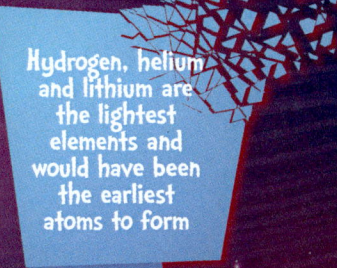

3 Lithium — Li

Atomic mass: 6.94
Electron configuration: 2, 1

> Hydrogen, helium and lithium are the lightest elements and would have been the earliest atoms to form

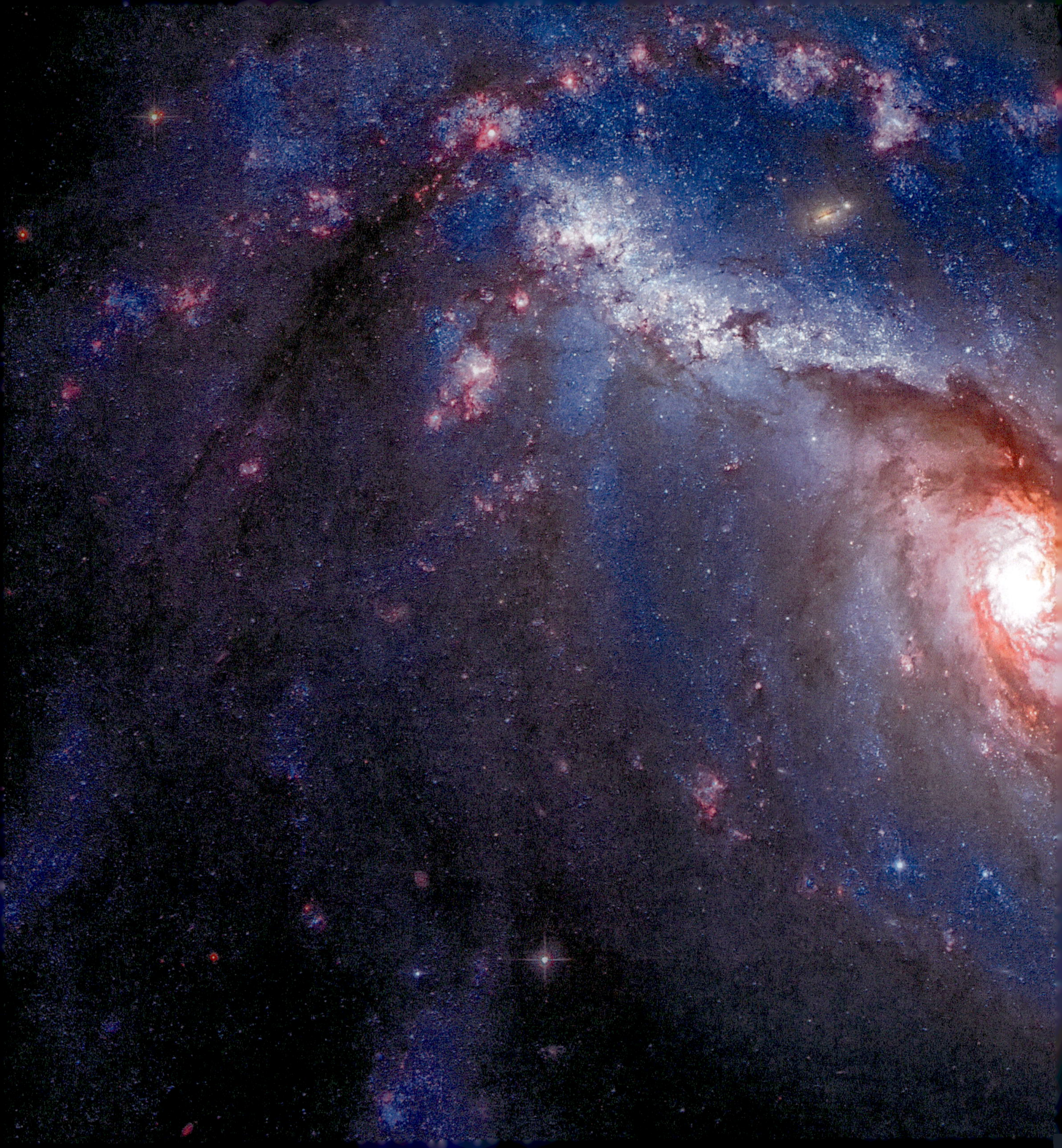

FORMATION OF STARS

A molecular cloud, sometimes called a stellar nursery if star formation is occurring within

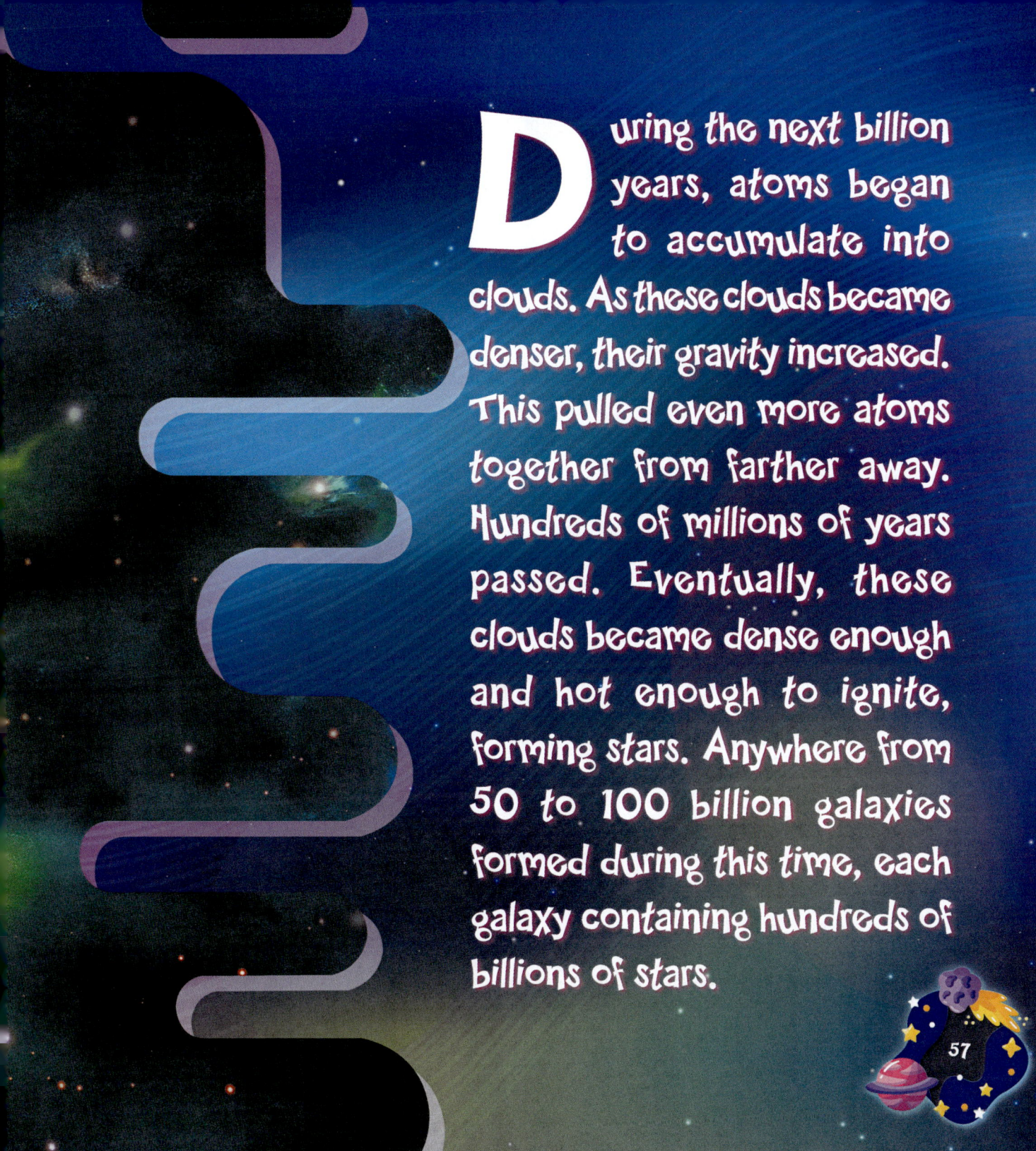

During the next billion years, atoms began to accumulate into clouds. As these clouds became denser, their gravity increased. This pulled even more atoms together from farther away. Hundreds of millions of years passed. Eventually, these clouds became dense enough and hot enough to ignite, forming stars. Anywhere from 50 to 100 billion galaxies formed during this time, each galaxy containing hundreds of billions of stars.

The first stars were comprised primarily of hydrogen and helium. As these stars carried on nuclear fusion, they created heavier elements. These elements are what make up planets, as well as life.

Heavier elements are what make up planets

As stars burned through their fuel supply, they became denser. Some stars eventually burned out and became red dwarves.

Red dwarf star

However, super high-mass stars became so dense that they collapsed and exploded in supernovae. These explosions scattered the heavy elements necessary for rock to form.

An illustration of a supernova

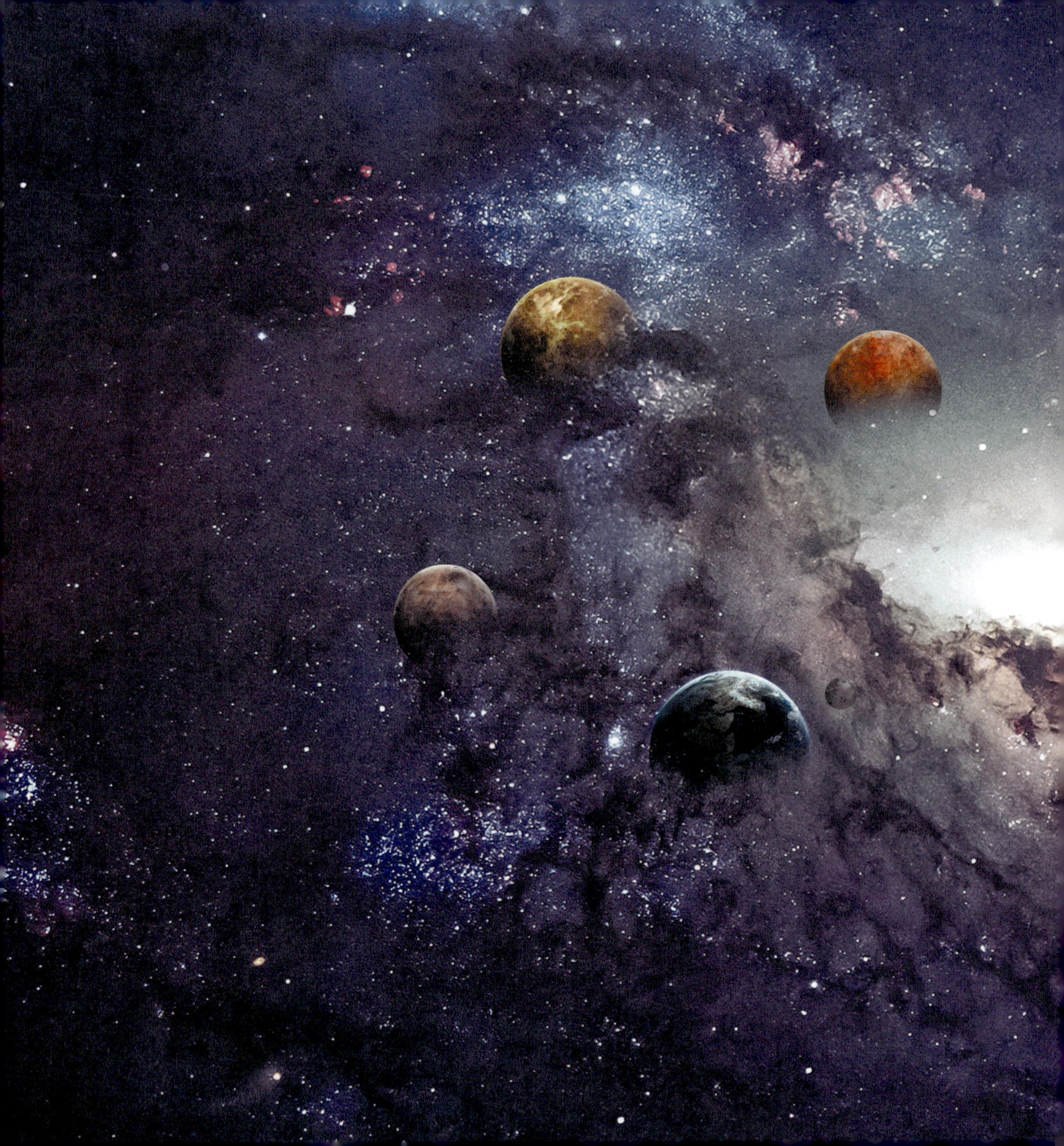

FORMATION OF PLANETS

Artist's concept of a protoplanetary disk, where particles of dust and grit collide and accrete forming planets or asteroids

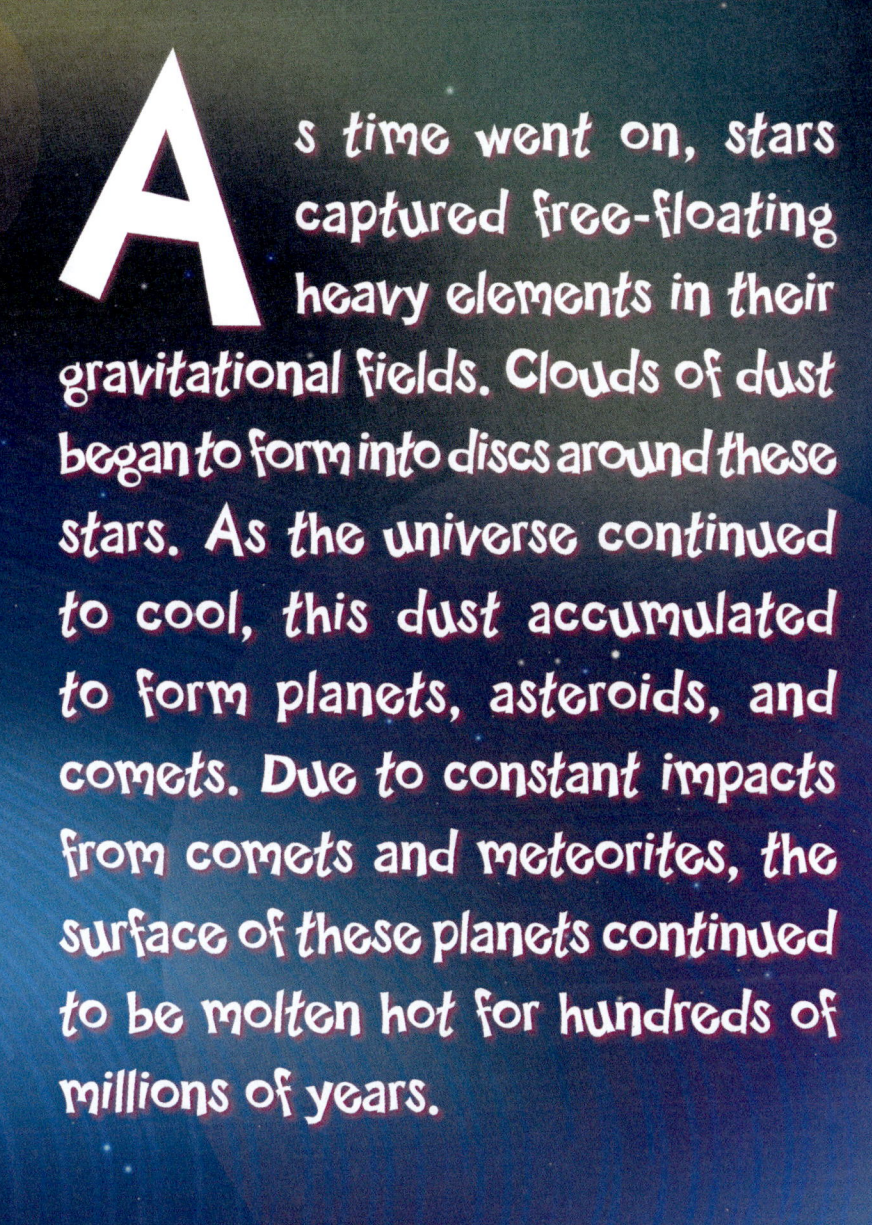

As time went on, stars captured free-floating heavy elements in their gravitational fields. Clouds of dust began to form into discs around these stars. As the universe continued to cool, this dust accumulated to form planets, asteroids, and comets. Due to constant impacts from comets and meteorites, the surface of these planets continued to be molten hot for hundreds of millions of years.

An illustration of the surface of a rocky planet

As the space within each solar system began to clear of debris, the surface of its rocky planets cooled and hardened.

SUMMARY

The universe, or cosmos, is everything that has existed, does exist, or will exist. It is believed to be about 13.8 billion years old. The observable universe is roughly 93 billion light years across, while the universe as a whole is constantly expanding and possibly infinite.

The Steady State Theory holds that the universe has no beginning or end and expands forever while maintaining a relatively constant density.

However, the Big Bang Theory is the most widely accepted explanation of how the universe began. It states that all of the contents of the universe were once super-dense and superheated. As this mass cooled and expanded, matter was formed. Light escaped and traveled away to form the Cosmic Microwave Background and atoms formed. These atoms, mostly hydrogen and helium, coalesced into stars and galaxies. Some massive stars created heavier elements through nuclear fusion and exploded, scattering these elements.

Stars collected dust into discs which condensed over time to form planets, asteroids, and comets. As cosmic debris cleared from solar systems, the surface of their rocky planets cooled and solidified.

The Big Bang Theory allows for the possibility that the universe will expand infinitely. However, a derivative of this theory states that the universe goes through a cycle of expansion and contraction at 80-billion-year intervals.

The universe gets bigger each day, and as technology advances, we will continue to see farther into the past and future.

VISIT

www.SpeedyBookStore.com

To view and download free content on your favorite subject and browse our catalog of new and exciting books for readers of all ages.

Made in the USA
Las Vegas, NV
03 April 2024